CINNAMON

by Amy C. Rea

Cody Koala

An Imprint of Pop!
popbooksonline.com

abdobooks.com

Published by Pop!, a division of ABDO, PO Box 398166, Minneapolis, Minnesota 55439. Copyright © 2022 by Abdo Consulting Group, Inc. International copyrights reserved in all countries. No part of this book may be reproduced in any form without written permission from the publisher. Cody Koala™ is a trademark and logo of Pop.

Printed in the United States of America, North Mankato, Minnesota.

052021
092021

THIS BOOK CONTAINS RECYCLED MATERIALS

Cover Photos: Shutterstock Images, foreground; iStockphoto, background
Interior Photos: Shutterstock Images, 1 (foreground), 7 (bottom right), 17, 21; iStockphoto, 1 (background), 5, 7 (bottom left), 9, 13, 18; Red Line Editorial, 7 (top); Oleksandr Rupeta/Alamy, 11; foodfolio/Alamy, 12; Vito Arcomano/Alamy, 14–15

Editor: Aubrey Zalewski
Series Designers: Laura Graphenteen and Colleen McLaren

Library of Congress Control Number: 2020948837
Publisher's Cataloging-in-Publication Data
Names: Rea, Amy C., author.
Title: Cinnamon / by Amy C. Rea
Description: Minneapolis, Minnesota : Pop!, 2022 | Series: How foods grow | Includes online resources and index.
Identifiers: ISBN 9781532169786 (lib. bdg.) | ISBN 9781098240714 (ebook)
Subjects: LCSH: Cinnamon--Juvenile literature. | Cinnamon tree--Juvenile literature. | Spice plants--Juvenile literature. | Agriculture--Juvenile literature. | Food crops--Juvenile literature.
Classification: DDC 631.5--dc23

Hello! My name is

Cody Koala

Pop open this book and you'll find QR codes like this one, loaded with information, so you can learn even more!

Scan this code* and others like it while you read, or visit the website below to make this book pop.

popbooksonline.com/cinnamon

*Scanning QR codes requires a web-enabled smart device with a QR code reader app and a camera.

Table of Contents

Chapter 1

What Is Cinnamon?

Cinnamon is a brown spice. It comes from the bark of cinnamon trees. There are two main types of cinnamon trees. The trees grow in **tropical** places.

Watch a video here!

Growing Cinnamon

Cinnamon trees come from Asia. It is warm there. The trees also grow in other parts of the world. They do well in the jungle. Jungles have the right kind of **climate**.

Where Cinnamon Grows

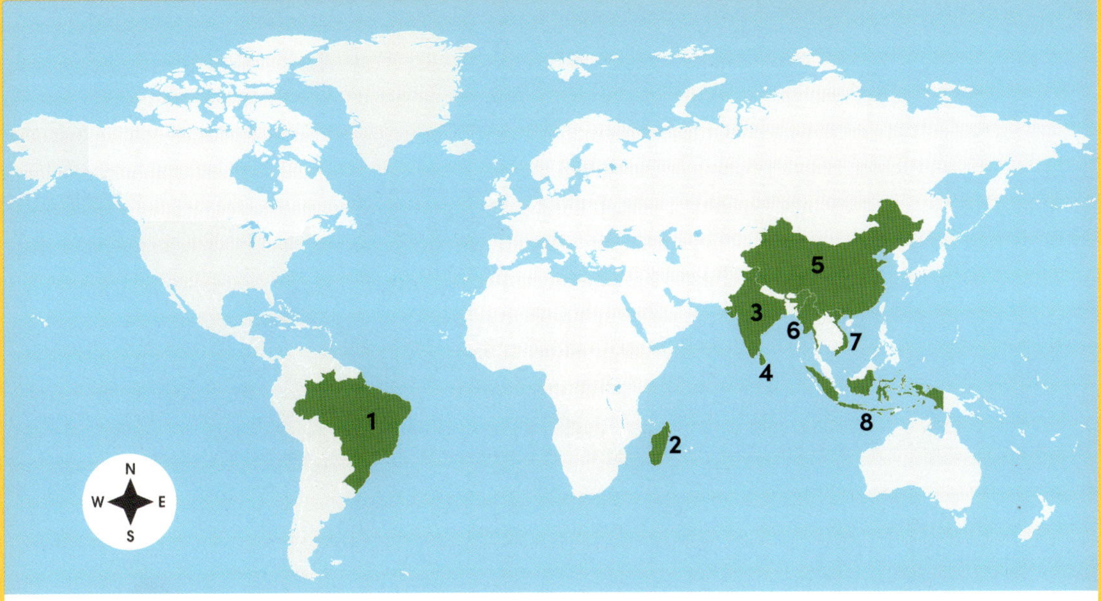

1. Brazil
2. Madagascar
3. India
4. Sri Lanka
5. China
6. Myanmar
7. Vietnam
8. Indonesia

Learn more here!

Cinnamon trees need warm temperatures to grow. They do well in 80 degrees Fahrenheit (27°C). The trees need lots of rain and damp air.

Some cinnamon trees can grow approximately 50 feet (15 m) tall.

Chapter 3

Harvesting Cinnamon

Cinnamon trees grow for at least two years. Then farmers cut the trees down. They leave the **stumps** and cover them with soil. The trees grow new **shoots**.

Complete an
activity here!

Cinnamon trees have
an outer and inner bark.
The outer bark is hard.

The inner bark is the spice.

It is softer and light brown.

The bark has a strong **aroma**.

Farmers cut down cinnamon trees to **harvest** the bark. Sometimes they cut down entire trees. Other times they cut only the branches.

Preparing Cinnamon

Farmers use hand tools to remove the trees' bark. They scrape away the outer bark. Then farmers peel off the inner bark with a sharp knife.

Learn more here!

Farmers lay out the bark strips. The cinnamon dries in the sun for several days. The strips curl and harden as they dry. The curled cinnamon is called a quill.

People cut the quills into smaller sizes. Stores can sell these whole. Mills grind other quills into powder. Cinnamon quills and powder flavor many foods around the world.

Cinnamon quills can flavor teas and stews. People often use cinnamon powder when they bake.

Making Connections

Text-to-Self

Have you ever tasted cinnamon? If so, what did it taste like? If not, would you like to try it?

Text-to-Text

Have you read books about another spice? How is that spice similar to cinnamon? How is it different?

Text-to-World

Cinnamon trees grow in tropical climates. Can you think of another food that grows in places like that?

Glossary

aroma – a strong and usually pleasing smell.

climate – the typical weather of a place or area over time.

harvest – to gather or pick crops.

shoot – a new growth on a plant.

stump – the part of a tree that stays in the ground after it has been cut.

tropical – having to do with a place where the weather is usually warm and wet.

Index

Online Resources

popbooksonline.com

Thanks for reading this Cody Koala book!

Scan this code* and others like it in this book, or visit the website below to make this book pop!

popbooksonline.com/cinnamon

*Scanning QR codes requires a web-enabled smart device with a QR code reader app and a camera.